"Science is a way of thinking much more than it is a body of knowledge."
-Carl Sagan

Dedicated to my mom, dad and sister who give me their continual support.

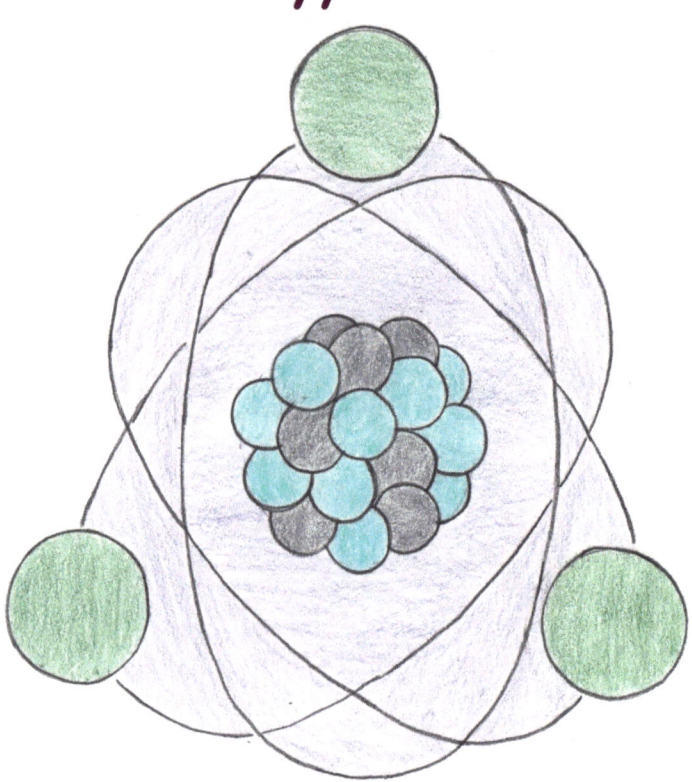

Text and Illustration Copyright © Talha Khalid

All rights reserved. No part of this book may be reproduced or utilized in any form or by any means electronic or physical, including photocopying, recording or any other information storage and retrieval system or device without the expressed written consent of Talha Khalid.

Printed by CreateSpace.com, an Amazon Company

LUKE AND THE JOURNEY THROUGH SCIENCE

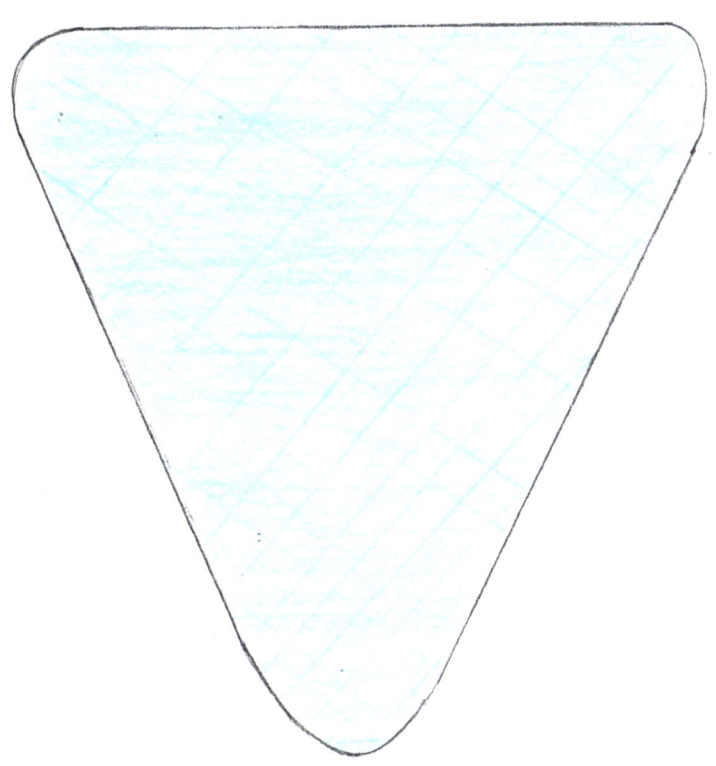

WRITTEN AND ILLUSTRATED BY TALHA KHALID

One bright morning, Luke walked to school. On his way to Hadron: The School of Science, Luke overheard some kids on the playground say, "Science isn't cool. It's only for nerds."

Luke ran to his teacher and yelled, "Teacher, teacher! The kids outside just said that science isn't cool." "Pay them no mind. Science is the coolest subject you'll ever study!" responded the teacher.

"You don't believe me, do you Luke? Come with me through the portal of history and I'll show you how science and technology has changed our world!"

The teacher walked Luke towards the portal through history.

"Let's start at a time when great minds journeyed into the world of science. A scientist by the name of Benjamin Franklin experimented with electricity while flying his kite. With his experiments and the help of other scientists, the nature of electricity was discovered and put to good use."

"How was electricity used?" asked Luke.

"Let's meet Thomas Alva Edison, the inventor of the lightbulb. The lightbulb uses electricity to light our houses. It's one of the greatest inventions of our lifetime!" said the teacher.

"Is that all electricity does?" asked Luke in a disappointed voice.

"You have much to learn." the teacher responded.

"Watch out!" Luke screamed as a lightning bolt almost struck him.

"This is the lab of Nikola Tesla, one of the greatest inventors of our time! Thanks to him, we have the alternating current system that powers everything, from our laptops to our TVs to our houses. He helped power our world!" the teacher said.

"That's great, but all this travel is making me hungry." said Luke. His stomach growled.

"How about some sweet potatoes and peanuts?" asked the teacher. "Yum, sounds delicious! Whose lab are we in now?" asked a startled Luke.

"This is the lab of George Washington Carver, a well-known scientist who helped farmers grow all kinds of crops. Thanks to his research and his passion for helping others, farmers could grow many new crops to support themselves and provide others with food." explained the teacher.

"Let's see what else science has done for us." as the teacher and Luke left the lab.

"I'm tired. How long do we have to keep walking?" asked Luke. Suddenly, they heard a loud "beep-beep."

"Look, someone is coming in a car." Luke said in excitement. "That's Henry Ford, one of the pioneers of the automobile. Thanks to him and many others like him, we can produce cars for all of us to drive. They can take us farther than we could ever go before!" explained the teacher.

"Wow, where else can technology take us?" asked Luke.

Suddenly, he and the teacher were on a plane. "How did we get up here?" he yelled.

"Thanks to the Wright Brothers, Orville and Wilbur, we can fly. After many failed attempts, they built a plane that can fly high and allow us to see the world from above" explained the teacher.

"Teacher, how high do you think we can go?" asked Luke excitedly.

"Is this how high you were thinking?" asked the teacher. "Are we in outer space?" exclaimed Luke.

"Yes, we're on the moon. Thanks to the engineers at the National Aeronautical and Space Administration, or NASA, we can travel to the moon. Space travel encouraged many to pursue careers in science and technology, including young students like yourself." They walked back into the rocket. "Time to go back to earth." said the teacher.

"Teacher, how does the rocket know to take us back home?" asked a bewildered Luke.

"Computers! We will tell them what to do!" the teacher exclaimed. "Computer, take us home!" commanded Luke.

"No, not like that! We have to program it." the teacher responded. "What's that?" asked a confused Luke.

"Programming means to give instructions to a machine or device, like computers on a rocket or even small computers like the one in your cellphone.

One of the first programmers was a woman by the name of Ada Lovelace. She wrote instructions, also known as code, to program machines. She paved the way for many future scientists, engineers and computer programmers.

Let's see how her work helped shape the future." said the teacher.

"Bing! You've got mail." "Whoa, are we on the computer screen? This looks so strange!" exclaimed Luke.

"Scientists, inventors and dreamers gave rise to the World Wide Web. With it, we can do many things, like connect, communicate and share information over the internet."

"That's great, but are we going to be on the screen forever?" asked Luke.

"We can leave. Just put on these glasses." said the teacher.

"RUN!" screamed Luke.

"Ha-ha, don't worry, Luke. This is virtual reality." explained the teacher. "Thanks to advances in computer technology, we can now feel like we're in another place entirely, without actually being there."

"So, none of this is real?" asked Luke. "Not at all. Try to touch the dinosaur." the teacher challenged Luke. "Hey, I can't touch it. Not fair!" protested Luke.

"Alright, time to take off the glasses and return to the classroom." the teacher said.

"You see, Luke, science and technology have led us to places we could never have imagined! It has helped the entire world move forward, and this is only the start. Who knows where we'll go from here? Maybe we'll have flying cars, new forms of energy and cures for all the world's diseases."

"And one day, Luke, you will be a great scientist! You, and many other dreamers like you, will use your creativity to help solve some of the world's biggest problems. Where science goes next is limited only by your imagination.

Are you ready to take a journey through science?"

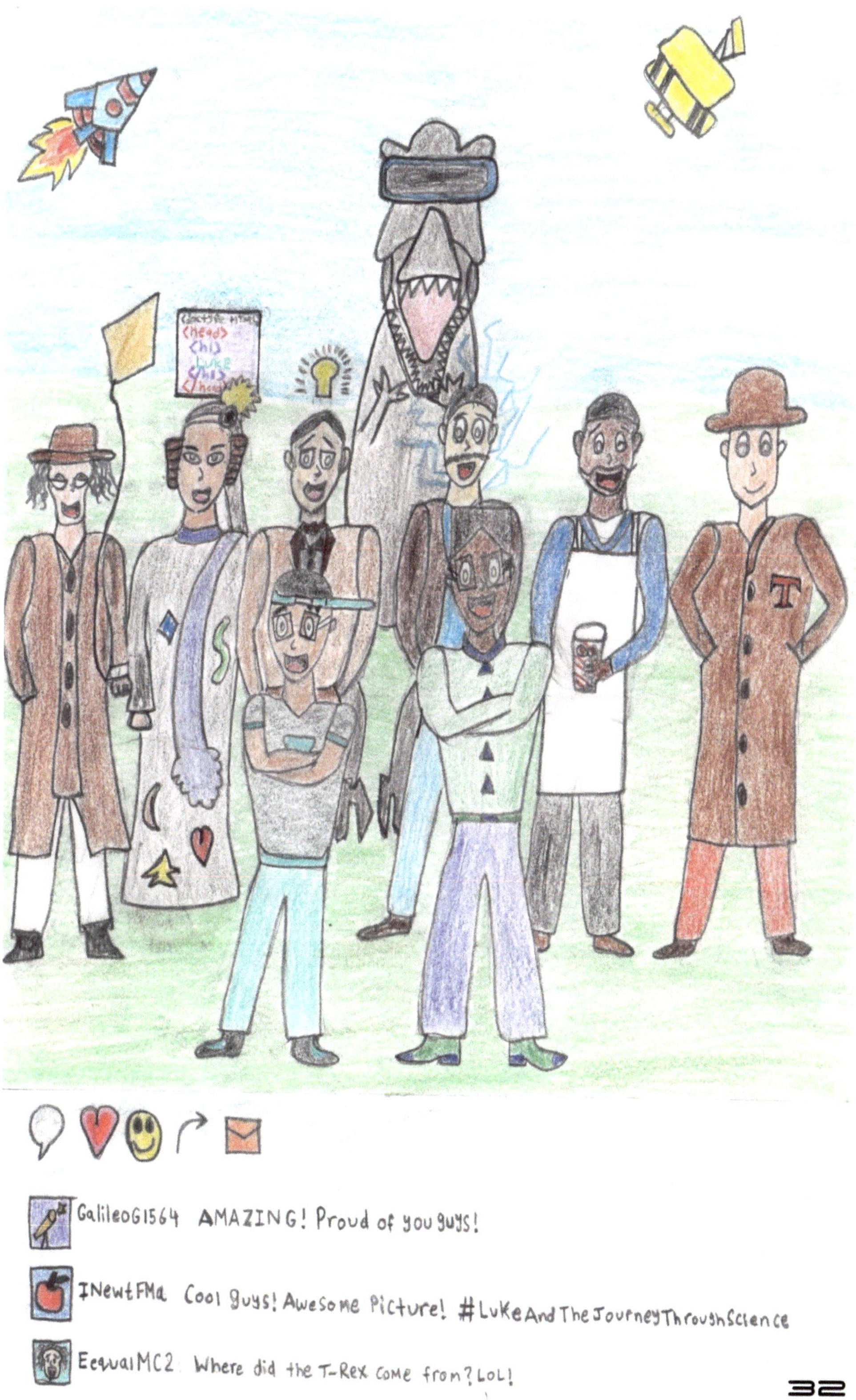

 GalileoG1564 AMAZING! Proud of you guys!

 INewtFMa Cool guys! Awesome Picture! #LukeAndTheJourneyThroughScience

 EequalMC2 Where did the T-Rex come from? LOL!

About the Author

Talha Khalid is a resident of Greenville, SC. Having graduated from Clemson University in 2016 with a Bachelor's of Science degree in Mechanical Engineering, he currently works as an engineer at a manufacturing facility in South Carolina. His passion for education, especially scientific education, inspires him to create content to advance learning amongst society's youth. He seeks to show the joys of science and technology to children, teens and adults.

CAN YOU FIND ALL 10 WORDS IN THIS WORD SEARCH PUZZLE?

```
W  Q  P  L  A  N  E  L  R  G
X  C  R  V  F  G  A  U  H  R
M  O  O  N  I  U  Y  K  T  O
J  K  G  D  T  D  S  E  C  C
Q  W  R  H  E  E  G  S  V  K
Z  D  A  V  F  G  S  J  M  E
J  B  M  D  V  B  U  L  B  T
Q  P  E  A  N  U  T  S  A  G
A  U  T  O  M  O  B  I  L  E
L  O  W  P  E  R  T  Y  Q  M
```

CODE   PROGRAM   MOON

PLANE   LUKE   ROCKET

TESLA   AUTOMOBILE   BULB

PEANUTS

www.ingramcontent.com/pod-product-compliance
Lightning Source LLC
Chambersburg PA
CBHW041943240526
45473CB00033B/497